My First BIG Book of ANIMAL LIFE CYCLES

by Ruth Owen

Ruby Tuesday Books

Published in 2025 by Ruby Tuesday Books Ltd.

Copyright © 2025 Ruby Tuesday Books Ltd.

All rights reserved. No part of this publication may be reproduced in whole or in part, stored in any retrieval system, or transmitted in any form or by any means, electronic, mechanical, photocopying, recording, or otherwise, without written permission from the publisher.

Editor: Mark J. Sachner
Design & Production: Emma Randall

Photo credits: Alamy: 39B (imageBROKER.com), 47T (Arterra Picture Library), 49B (Joe Blossom), 54T (All Canada Photos), 55 (Chris Mattison), 61T (Bert de Ruiter), 63T (WILDLIFE GmbH), 71B (Brandon Cole Marine Photography), 76 (Derek Mitchell); Stefan Christmann (Nature Picture Library): 40–41, 42; Corbis: 17B; Katherine Feng (Nature Picture Library): 16B, 18; FLPA: 36T, 39T, 46, 48B, 49T; Getty Images: 37T; Eric Isselee (Shutterstock): 4B, 5T, 12T, 26T, 93B; iStock: 33T (wwing); Macrovector (Shutterstock): 91, 92T; Nature Picture Library: 7B (Ingo Arndt), 14T (Nick Upton), 14B (Jose Luis Gomez de Francisco), 20T (Jane Burton), 21T (Klein & Hubert), 21B (Klein & Hubert), 22T (Juergen Freund), 25B (Tony Wu), 27T (David Tipling), 28B (Steven David Miller), 29C (Ingo Arndt), 34 (Folkert Christoffers), 35B (Folkert Christoffers), 45B (Oliver Richter), 50 (Remi Masson), 51T (Sinclair Stammers), 52T (John Cancalosi), 52B (Jack Dykinga), 53T (Rolf Nussbaumer), 53B (John Cancalosi), 60T (Sergey Gorshkov), 66T (Solvin Zankl), 67B (Juergen Freund), 68 (Michel Roggo), 69T (Kim Taylor), 72–73 (Fred Bavendam), 79T (Rolf Nussbaumer), 82 (Ingo Arndt), 85T (Heidi & Hans-Juergen Koch), 85B (Laurent Geslin); Science Photo Library: 89 (Claude Nuridsany & Marie Perennou); Shutterstock: Cover (chinahbzyg/Ermolaev Alexander/Roger ARPS BPE1 CPAGB/ BranoMolnar), 4T, 5B, 6, 7TL, 7TR, 7C, 8T, 8B, 9L, 9R, 10, 11T, 12B, 13, 15, 16T, 19, 20B, 21C, 22B, 25T, 26, 27, 28T, 29, 30, 31, 32, 33, 35T, 36B, 37B, 38, 40, 43, 44T, 47B, 48T, 51C, 51B, 54B, 56T, 57, 58, 59T, 61B, 62, 63B, 64T, 65, 66, 67T, 67C, 69B, 70T, 71T, 74, 75, 77T, 77C, 78, 79B, 80, 81, 84–85, 84T, 86, 87, 88, 90, 92B, 93T (for a full detailed list of contributors, contact info@ rubytuesdaybooks.com); Superstock: 17T (Katherine Feng/Minden Pictures), 23T (Minden Pictures), 23B (Anup Shah/Minden Pictures), 24 (Hiroya Minakuchi/Minden Pictures), 56 (Christian Ziegler), 60 (IMAGO), 77B (Andrew McLachlan), 83C (Roger Tidman); Vishnevskiy Vasily (Shutterstock): 26T, 44B, 45; Warren Photographic: 11B.

British Library Cataloguing in Publication Data (CIP) is available for this title.

ISBN 978-1-78856-582-0

Printed in Malta by Gutenberg Press Ltd.

www.rubytuesdaybooks.com

What's Inside?

What Is a Life Cycle?..................................4

Are All Animal Life Cycles the Same? 6

Which Baby Animals Drink Milk? 8

How Many Babies Do Mammals Have? 10

Do All Newborn Mammals Act the Same?..12

Do All Baby Mammals Look Like Their Mum?..14

Which Baby Mammal Fits in Mum's Paw? 16

How Does a Panda Cub Change and Grow? 18

Which Baby Mammal Has Spines? 20

Which Baby Mammal Must Go to School?...22

Which Baby Is the Biggest of All? 24

How Do Bird Life Cycles Begin? 26

Which Bird Eggs Are the Smallest and the Biggest? 28

Do All Baby Birds Look the Same?..........30

How Does a Blackbird's Life Cycle Begin? ...32

How Do Blackbird Chicks Grow
and Change? 34

Where Does a Swan Lay Her Eggs? 36

When Can a Swan Cygnet Swim? 38

Which Egg Keeps Warm on
Dad's Feet? 40

What Does a Penguin Chick Eat? 42

Which Baby Bird Is Bigger Than
Its Parents? 44

How Many Eggs Does a Frog Lay? 46

How Does a Frog Tadpole Change? 48

Who Lays Long Strings of Eggs? 50

Which Amphibian Starts Life
in a Puddle? 52

Why Must Spadefoot Toads Grow
Up Fast? 54

Do Baby Lizards Hatch from Eggs? 56

How Do Baby Iguanas Keep Safe? 58

Which Reptile Mum Guards Her Eggs? ... 60

How Does a Croc Mum Help
Her Babies? 62

Who Digs a Nest on a Beach? 64

Why Must Baby Turtles Run? 66

Do Fish Lay Eggs? 68

Which Caring Mum Has Eight
Long Tentacles? 70

How Do Octopus Mums Care for
Their Eggs? 72

Which Animal Mum Might Eat Dad? 74

How Does a Spider Mum Protect
Her Eggs? 76

Which Baby Is Called a Caterpillar? 78

How Does a Caterpillar Become
a Butterfly? 80

Which Animal Life Cycle Begins
in a Hive? 82

What Do Baby Honeybees Eat? 84

How Many Eggs Does a Ladybird Lay? .. 86

How Do Baby Ladybirds Become
Red and Spotted? 88

What Goes Around and Around? 90

Around, and Around, and Around! 92

My Life Cycle Words 94

Big Life Cycles Quiz 96

What Is a Life Cycle?

A life cycle is the story of all the stages in the life of a living thing.

Animals have four main stages in their lives.

Stage 1
An animal is born, and its life begins.

A three-week-old lion cub

Stage 2
An animal grows bigger. Sometimes the way it looks changes.

3 months old

9 months old

4 years old

Stage 3

An animal grows up and becomes an adult. Now it is ready to **reproduce** and have babies of its own.

Adult male lion 8 years old

Adult female lioness 8 years old

Cubs

Stage 4

An animal gets older, and one day it dies.

Lions live for about 10 to 14 years.

This lion is 13, which means he is very old.

Are All Animal Life Cycles the Same?

The ways in which animals are born, grow and reproduce can be very different.

Some animals grow inside their mother's body.

There are kittens growing inside this mother cat.

A baby chicken

Some animals grow inside an egg.

Some animals go through **BIG** changes as they become an adult.

Young puss moth

Adult puss moth

Mother elephant

A female elephant is ready to have a baby when she is about 15 years old.

Baby aphid

Mother aphid

A tiny female aphid can have babies when she is just one week old!

Which Baby Animals Drink Milk?

We sort animals into different groups like the ones below.

The way an animal looks and how it lives tells us which group it belongs to.

Mammals   Birds   Fish   Amphibians

Reptiles   Insects   Spiders

Mammals are animals with fur or hair.

They breathe air with body parts called **lungs**.

Hair

Breathing air

Giraffe

Female mammals give birth to live babies.

Here comes baby!

Mum pushes hard.

The calf is born.

Hello, Baby!

Mother giraffe

Calf drinking milk

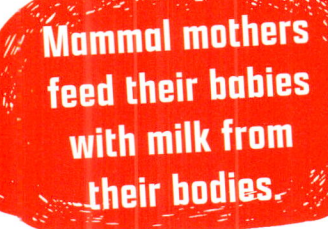
Mammal mothers feed their babies with milk from their bodies.

How Many Babies Do Mammals Have?

Some mammal mothers give birth to one baby at a time.

An elephant may have one calf about every five years.

A calf drinking milk

It takes 22 months for an elephant calf to grow inside its mum!

This mother sheep gave birth to twin lambs.

Mother sheep

Lamb

Mother cat

Kittens drinking milk

Other mammal mums, such as dogs, cats, rabbits and mice may have lots of babies at once.

It took just nine weeks for all these pups to grow inside their mum!

Father dog

Mother dog

Puppies feeding

Do All Newborn Mammals Act the Same?

All newborn mammals drink milk, but they can be different in other ways.

Newborn tiger cubs cannot see or walk.

Six-day-old tiger cubs

They open their eyes when they are about six days old.

A mother tiger carries her babies in her mouth.

A tiger cub can walk when it is about four weeks old.

Mother moose

A newborn moose can walk after just two hours.

Calf

This little calf was born one hour ago.

After two days, a moose calf can run with its mum – fast!

It may need to run away from hungry bears and wolves.

Do All Baby Mammals Look Like Their Mum?

No! Some newborn baby mammals look very different than their mums.

A wood mouse gives birth to her babies in an underground nest.

Mother wood mouse

Nest hole

Newborn wood mouse

The newborn mice have no fur and cannot see.

Seven-day-old mouse

After one week, the baby mice start to grow fur.

After two weeks, their eyes open.

When a baby wood mouse is three weeks old, it looks like its mum.

Now it is ready to begin its adult life!

Three-week-old mouse

Which Baby Mammal Fits in Mum's Paw?

In a **den** in a forest, a tiny panda cub is born.

Panda den

The cub's eyes are closed and she cannot see.

Mother panda

Two-day-old cub

Her pink skin is covered with just a little fuzzy white fur.

A mother panda gently holds her baby in her mouth or paw.

The panda cub drinks milk from her mother's body.

At two weeks old, the panda cub looks more like mum!

Two-week-old cub

Black eye patches

Black ears

Black legs

Turn the page to see the panda cub grow!

How Does a Panda Cub Change and Grow?

Soon the cub's eyes open and her thick fur grows.

At four months old, the cub leaves the den to play.

A mother panda licks her cub to keep it clean.

The cub learns to climb trees.

The cub tries eating tough, crunchy bamboo — just like her mum.

Six-month-old cub

At two years old, the young panda leaves her mum to live alone in the forest.

Two-year-old panda

She will be ready to have her own cub at six years old.

Which Baby Mammal Has Spines?

In spring, male and female hedgehogs meet up to **mate**.

Female hedgehog

Hoglet

After four weeks, a female hedgehog gives birth to five spiny hoglets in her nest.

The hoglets drink milk from their mother.

The hedgehog's nest is under this shed.

If a dog or fox finds her nest, a mother hedgehog carries her hoglets to a new safe place.

At four weeks old, the hoglets go hunting with their mum.

Hedgehog Food
- Worms
- Slugs
- Millipedes
- Snails
- Beetles

Mother hedgehog

Four-week-old hoglet

A hedgehog is grown-up at one year old. It lives for about five years.

Which Baby Mammal Must Go to School?

It's not just human children that go to school. Young orangutans have lessons, too!

Mother orangutan

A baby orangutan drinking

A newborn baby orangutan drinks milk and holds onto its mum's hair.

Once the baby is two years old, mum shows it how to climb.

A mother orangutan teaches her baby which fruits, leaves and other foods are good to eat.

This baby is tasting chewed-up fruit.

The young orangutan learns how to make a sleeping nest from branches and leaves.

Nest

A young orangutan lives with its mum for up to nine years.

Which Baby Is the Biggest of All?

A newborn blue whale is 7 metres long. It weighs as much as an elephant!

A female blue whale is ready to reproduce when she is 10 years old.

Nine-month-old blue whale calf

Mother whale

The calf, or baby, is born underwater.

It quickly swims to the surface to breathe some air.

A blue whale calf drinks milk from its mum.

It drinks enough milk to fill two bathtubs each day!

When a calf is one year old, it is ready to live on its own.

A blue whale's tail

Scientists think that blue whales can live for 90 years.

How Do Bird Life Cycles Begin?

All baby **birds** begin their life cycle inside an egg.

Cassowary   Chicken   Cormorant   Puffin   Quail   Thrush   Sparrow

Before a mother bird lays her eggs, she must find or build a safe nest.

Storks build huge nests of twigs on chimneys and rooftops.

Mother stork

Nest

Father stork

Father birds often help build nests, too.

Robin chick

Nest

A pair of robins built this nest inside a cycle helmet!

A tawny owl lays her eggs inside a tree hole nest.

Father tawny owl

Mother tawny owl

Chick

The chicks are safe inside the tree hole.

Which Bird Eggs Are the Smallest and the Biggest?

Hummingbirds lay the smallest eggs of all!

Hummingbird

The eggs are just the size of peas or jellybeans.

Hummingbird eggs

Nest

Scientist's hand

This is a tiny hummingbird's nest.

It is made of leaves and grass, held together by threads from spiderwebs.

Ostrich egg

Chicken egg

A female ostrich lays the biggest eggs of all.

An ostrich egg can weigh about 1.5 kilograms.

It takes about 42 days for a chick to grow inside its egg.

An ostrich chick hatching

Mother ostrich

Ostrich chicks

Do All Baby Birds Look the Same?

No! Some baby birds are pink and bald when they hatch. Others have feathers.

Wagtail chick

This wagtail chick has no feathers. It cannot see, walk or fly.

Two-week-old wagtail chick

After one week, the chick's feathers grow and its eyes open.

At two weeks old, the chick is ready for flying lessons.

A duckling has sticky, wet feathers when it hatches.

Duckling

Dry, fluffy feathers

After just a few hours, this baby can walk.

When it's one day old, a duckling can swim with mum.

Mother duck

How Does a Blackbird's Life Cycle Begin?

In spring, a male and a female blackbird meet up to mate.

The female builds a nest from grass and twigs.

Grass

Female blackbird

Blackbird eggs

She lays her eggs inside the nest.

Female blackbird

Nest

The female blackbird sits on the eggs to keep them warm.

The warmth helps her chicks grow.

After two weeks, chicks hatch from the eggs.

blackbird chick

The tiny, bald blackbird chicks cannot see.

Now mum and dad must catch worms and **insects** for the chicks to eat.

How Do Blackbird Chicks Grow and Change?

The parent blackbirds work hard bringing the chicks food to eat.

The chicks soon open their eyes.

Their feathers start to grow.

Father blackbird

One-week-old chick

At 14 days old, the chicks leave the nest.

They flutter around and learn to fly.

A 14-day-old chick

The young blackbirds still beg their mum and dad for food.

Father blackbird

Chick

The chicks live with their parents for about six weeks.

At one year old, they can reproduce and have chicks of their own.

35

Where Does a Swan Lay Her Eggs?

Swans are large water birds that live on ponds, lakes and rivers.

A female swan is called a pen.

Nest

A male swan is called a cob.

In spring, a male and female swan build a huge nest of sticks and grass.

The female swan lays six eggs in the nest.

Swan eggs

The parent swans take it in turn to sit on the eggs to keep them warm.

Egg

After about five weeks, a baby swan hatches from each egg.

A baby swan is called a cygnet.

Cygnet

Let's say it! "SIG-nuht"

When Can a Swan Cygnet Swim?

When a cygnet is just two days old, it is ready to swim with its parents.

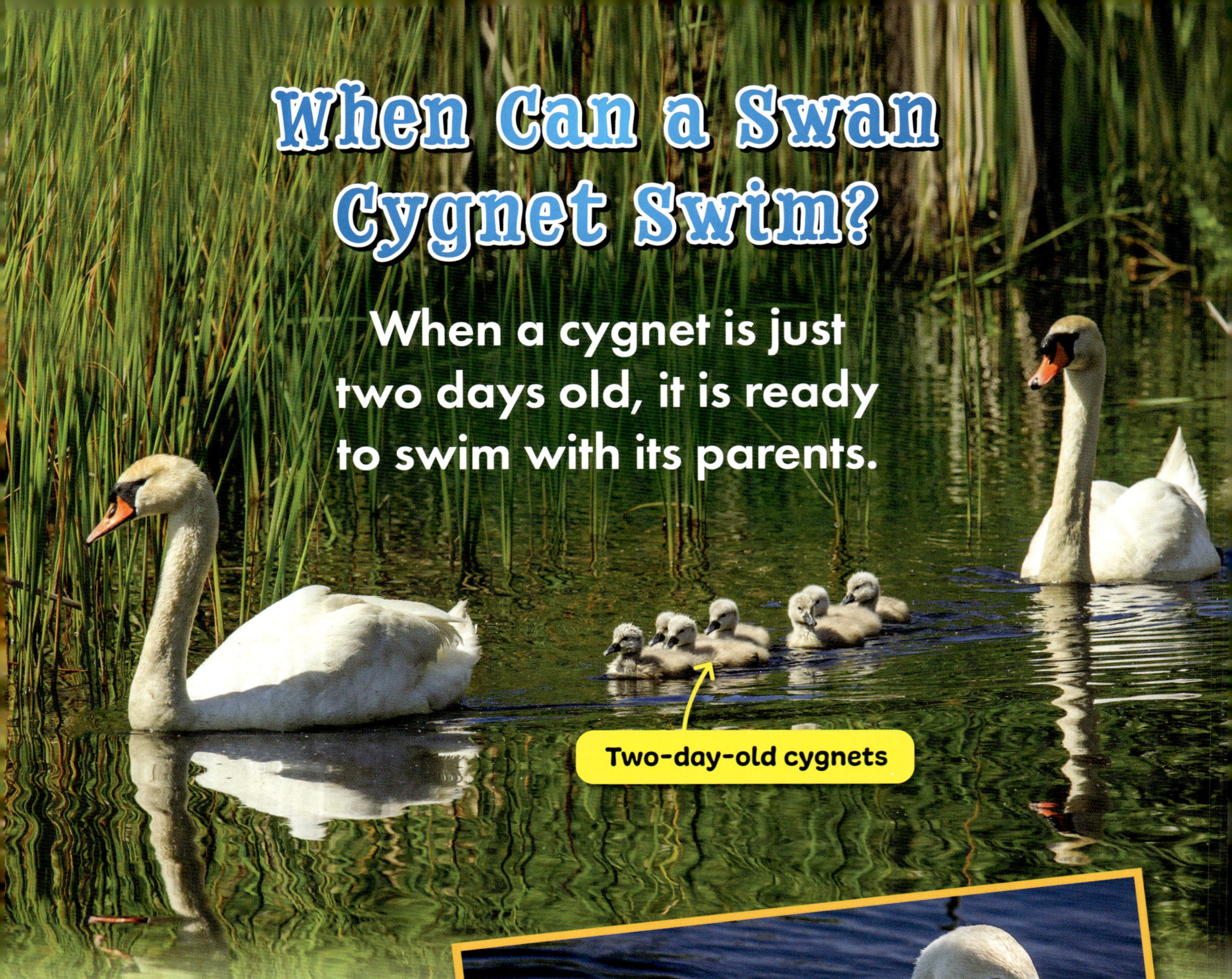

Two-day-old cygnets

If a cygnet gets tired, it rides on its mum or dad's back!

A cygnet gets bigger and grows brown feathers.

A three-month-old cygnet

At five months old, a young swan can fly.

The young swan starts to grow white feathers.

Now it is ready to leave its parents and live on its own.

Which Egg Keeps Warm on Dad's Feet?

Emperor penguins live in icy Antarctica.

North America

South America

Antarctica is down here.

A mother emperor penguin lays her egg on the cold ground.

Father penguin

She rolls it onto the father penguin's feet.

Egg

Mother penguin

Now the mother penguin must walk to the sea to hunt for **fish**.

The walk takes many days!

The father penguin keeps the egg warm and safe on his feet.

Lots of father penguins huddle together to keep warm.

Father penguins

Snow falls and icy winds blow!

After 10 weeks, a tiny chick hatches.

Chick's beak

Egg

Father's feet

What happens next?

What Does a Penguin Chick Eat?

The hungry penguin chick keeps warm on its father's feet.

Chick

Mother penguin

The mother penguin walks back from the ocean.

She has eaten lots of fish.

She spits up runny, fishy food for the chick.

Chick eating

As the chick gets bigger, both parent penguins go hunting.

The chick waits for food with other chicks.

The chick's adult feathers start to grow.

White adult feathers

Fluffy chick feathers

Soon it will be ready to walk to the sea and hunt fish for itself.

Which Baby Bird Is Bigger Than Its Parents?

It's a cuckoo chick!

Female cuckoo

A female cuckoo lays her egg in another bird's nest.

Then she flies away.

Cuckoo egg

Nest

The nest belongs to a tiny reed warbler bird.

The reed warbler sits on her eggs – and the cuckoo's egg.

After 12 days, the cuckoo hatches.

Female reed warbler

Cuckoo chick

One by one, the cuckoo chick pushes the other eggs out of the nest.

Now the parent reed warblers have just one big chick to feed.

The cuckoo grows bigger and bigger.

The parent birds don't realise that the huge chick is not their own!

Cuckoo chick
Parent reed warbler

How Many Eggs Does a Frog Lay?

A frog is an **amphibian**. Most amphibians lay their eggs in water.

Eggs

Female frog

In spring, a female common frog lays up to 2000 eggs in a pond.

Each tiny black egg is inside a round blob of soft, clear jelly.

After several weeks, an egg becomes a tadpole.

The tadpole wriggles out of the jelly.

A tadpole can breathe underwater.

The tadpole breathes with body parts called gills.

How Does a Frog Tadpole Change?

After about 16 weeks, a tadpole starts to change. First, it grows back legs.

Algae

Tadpole

Tadpoles eat slimy, green algae and tiny pond animals.

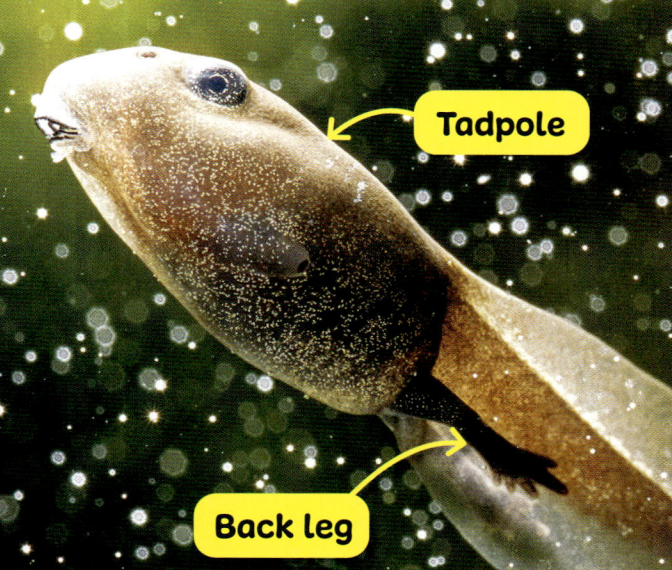

Tadpole

Back leg

Soon a tadpole's front legs grow.

Its tail gets shorter, until it has no tail.

It also grows lungs for breathing.

Shorter tail
Back leg
Front leg

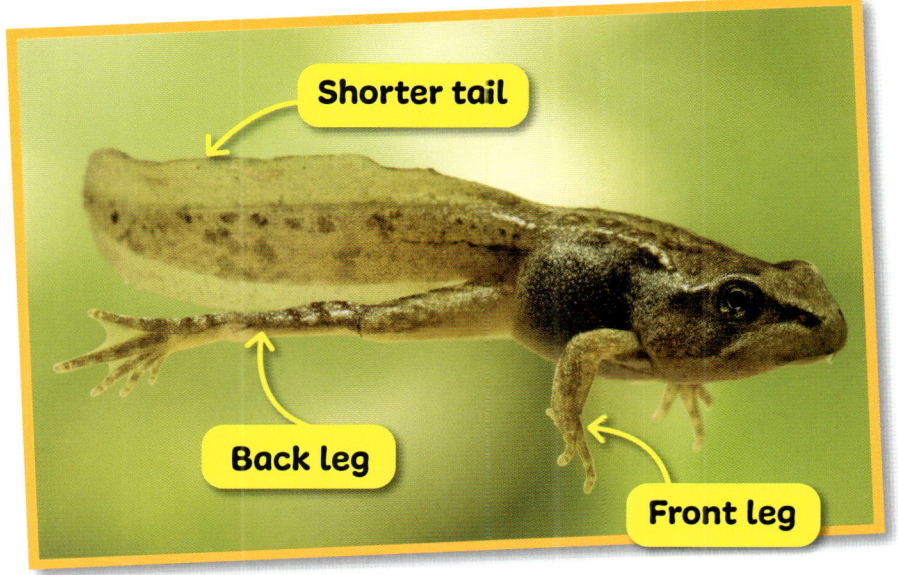

The tadpole becomes a froglet that lives on land and in water.

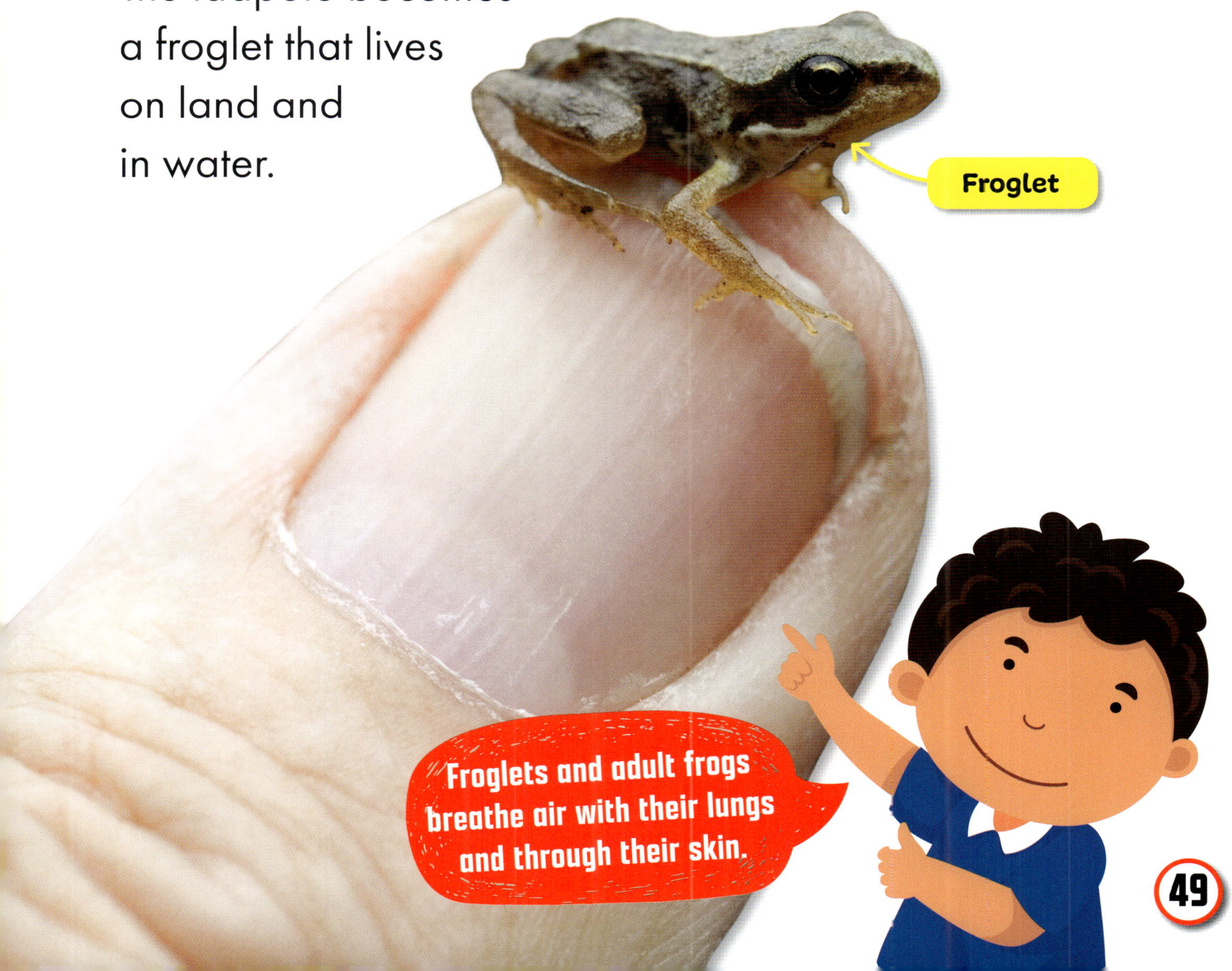
Froglet
Froglets and adult frogs breathe air with their lungs and through their skin.

Who Lays Long Strings of Eggs?

In spring, a male and female toad meet up in a pond to mate.

Then the female toad lays 5000 tiny eggs in strings.

Female toad

Toad eggs

Toads are amphibians. They start their life cycles in water.

Tadpoles hatch from the eggs and live in the pond.

The tadpoles grow back legs.

Then they grow front legs.

Their tails get shorter and shorter.

The tadpoles become toadlets that live on land.

Eight-week-old toadlet

Adult toad

A toad becomes an adult at four years old.

51

Which Amphibian Starts Life in a Puddle?

**It's a spadefoot toad!
These amphibians live in hot, dry deserts.**

They keep cool by hiding in underground **burrows**.

When rain falls, the toads leave their burrows.

Spadefoot toad

A desert storm

Spadefoot toads have feet that are a good shape for digging — like spades.

Vocal sac

Male spadefoot toad

The male toads call to females.

A male inflates his vocal sac to make his call louder.

Male and female toads mate in puddles.

Then each female lays up to 3000 eggs.

Tadpoles

In less than one day, tadpoles hatch!

Why Must Spadefoot Toads Grow Up Fast?

Day by day, the puddles of rain get smaller. They will soon dry up.

The spadefoot toad tadpoles must change fast.

Tadpole

They grow from tadpoles to toads in just 10 days!

Next, the little toads eat lots of ants, beetles, grasshoppers and **spiders**.

Do Baby Lizards Hatch from Eggs?

Reptiles are animals with **scaly** skin. Lizards, snakes and crocodiles are reptiles.

Green iguanas are large lizards.

Female green iguana

Spiny back

Burrow

A female iguana digs a burrow.

She lays about 30 eggs inside.

Iguana eggs

Iguana eggs have soft, leathery shells.

Iguana egg

Mother iguanas do not take care of their eggs or babies.

After about three months, a tiny green iguana hatches from each egg.

Baby green iguana

What happens next?

How Do Baby Iguanas Keep Safe?

Once baby green iguanas hatch, they quickly climb into trees.

The baby's green skin helps it hide among leaves.

This keeps it safe from hungry snakes and birds.

Snout

One-day-old iguana

Tail

Long toes with claws

This baby iguana is real-life size!

It is 20 centimetres long from its snout to the end of its tail.

The little iguanas eat leaves and flowers.

As an iguana gets older, spines grow on its back.

At four years old, a green iguana is an adult.

It measures 1.5 metres long!

Spines

Scales

Adult iguana

A green iguana can live for 20 years.

Which Reptile Mum Guards Her Eggs?

Crocodiles are the biggest of all the reptiles.

Mother crocodile

The nest is under here.

A mother crocodile digs a nest hole in sandy ground.

Crocodile eggs

She lays up to 80 eggs in the nest and covers them with soil.

Covering the eggs keeps them warm and helps the babies grow.

The mother crocodile guards her nest.

ROAR!

If an egg-eating **predator** comes close, she growls and attacks!

Baby crocodile

After 90 days, the baby crocodiles start to hatch.

How Does a Croc Mum Help Her Babies?

In their underground nest, the crocodile hatchlings call to their mum.

Mum uses her snout and claws to help her babies escape.

Crocodile hatchling

Egg

Now the babies must hide from predators such as snakes and wild dogs.

Mother croc

Hatchling

The mother croc gently carries her babies to a river in her huge mouth.

The hatchlings hide among plants in the shallow water.

Mum stays close to guard them!

A crocodile is grown-up at 12 years old.

It may live to be more than 100!

Who Digs a Nest on a Beach?

Green sea turtles are large reptiles that live in the ocean.

Male and female turtles meet up and mate in the sea.

Green sea turtle eating sea grass

When she's ready to lay eggs, a female turtle visits a beach.

She digs a large nest hole with her flippers.

Female turtle

Nest hole

Flipper

The female turtle lays more than 100 eggs in her nest.

Turtle eggs

Then she covers the eggs with sand.

The female turtle goes back to the ocean and does not take care of her eggs.

Why Must Baby Turtles Run?

After 60 days, tiny turtles hatch from the eggs.

Hatchling

The hatchlings dig up to the surface.

This hatchling's tiny shell is just 5 centimetres long.

Now seabirds and other predators will try to catch the hatchlings.

The babies must quickly run to the sea.

Then they swim away to begin their life in the ocean.

In about 25 years, the turtles will be adults.

The females will come back to this beach to lay their own eggs!

Do Fish Lay Eggs?

Yes! Most female fish lay eggs. But some fish give birth to live babies.

Rainbow trout reproduce in rivers.

The female fish lays 3000 tiny eggs on the stones.

Male

Female rainbow trout

Rainbow trout eggs

The male fish covers the eggs with a substance called milt.

Real-life size egg

Now tiny fish will grow in the eggs.

After four weeks, babies called fry hatch from the eggs.

The yolk sac gives the fry food.

Egg

Yolk sac

This rainbow trout fry is hatching.

This young trout is called a fingerling.

Six-month-old fingerling

That's because it's the size of an adult's finger.

Which Caring Mum Has Eight Long Tentacles?

It's an octopus mum.

A Giant Pacific octopus weighs the same as seven children!

Giant Pacific octopus

Eight arms, or tentacles

Suckers

After mating, a female giant octopus finds a crack in a rock.

This will be her den.

The octopus lays up to 400,000 tiny eggs.

The eggs come out of her siphon, or funnel.

Siphon

She uses her suckers to plait the eggs into chains.

Chains of eggs

It takes 40 days to lay all those eggs!

Then she sticks the chains to the roof of the den.

How Do Octopus Mums Care for Their Eggs?

A mother Giant Pacific octopus guards her eggs from hungry predators.

She whooshes water from her siphon over the eggs to keep them clean.

She picks off little bits of dirt with her suckers.

Chains of eggs

Rock den

Mother octopus

After six months, tiny **larvae** hatch from the eggs and swim away.

Baby octopus, or larva

While she guards her eggs, a mother octopus doesn't eat.

Egg

Hatching larva

Once the babies hatch, the mother octopus dies.

Giant Pacific octopuses live for three to five years.

Which Animal Mum Might Eat Dad?

Female garden spiders build webs with silk from their bodies.

Female garden spider

Flies and other insects get trapped in the web.

Then the spider eats them.

Female spider

Trapped wasps

In summer, it's time for garden spiders to mate.

A male spider gently plucks the silk threads of a female's web.

This tells the female he wants to mate and is not food.

Male spider

Female spider

After mating, the male spider quickly escapes.

The female might eat him!

How Does a Spider Mum Protect Her Eggs?

After mating, a female garden spider lays up to 800 eggs.

She wraps the eggs in an egg sac made of silk.

Mother spider

Silk egg sac

The mother spider stays close to guard her eggs.

When winter comes, the mother spider dies.

Ball of spiderlings

In spring, tiny spiderlings hatch.

They leave the egg sac and stick close together in a ball.

After one week, the spiderlings start to build their own tiny webs.

Spiderling

Adult garden spider

At two years old, the spiderlings will be ready to reproduce.

Which Baby Is Called a Caterpillar?

**It's a butterfly baby!
This is the life cycle of a monarch butterfly.**

A butterfly is a type of insect. Insects are tiny animals with six legs.

Butterfly laying eggs

Egg

After mating, a female monarch butterfly lays about 400 eggs on leaves.

After four days, a tiny caterpillar hatches from each egg.

The caterpillars eat and eat.

This caterpillar is eating its egg case.

As a caterpillar grows bigger, its skin gets too tight!

Old skin

1

2

New skin

3

It squeezes out of the small skin, and there's a new, bigger skin underneath.

How Does a Caterpillar Become a Butterfly?

After about 14 days, something amazing happens.

A monarch butterfly caterpillar hangs upside down.

Caterpillar

Suddenly, its skin splits open.

Split skin

Caterpillar skin

Chrysalis

Inside its caterpillar skin, the insect has become a chrysalis.

Which Animal Life Cycle Begins in a Hive?

It's a honeybee's life cycle!

Bees belong to the insects group of animals.

Honeybees live in a home called a hive.

Honeybees

Hive entrance

About 50,000 worker bees and a queen bee live in the hive.

Their hive is inside a tree trunk.

Inside the hive are combs made of six-sided holes called cells.

The queen bee lays a tiny egg in each cell.

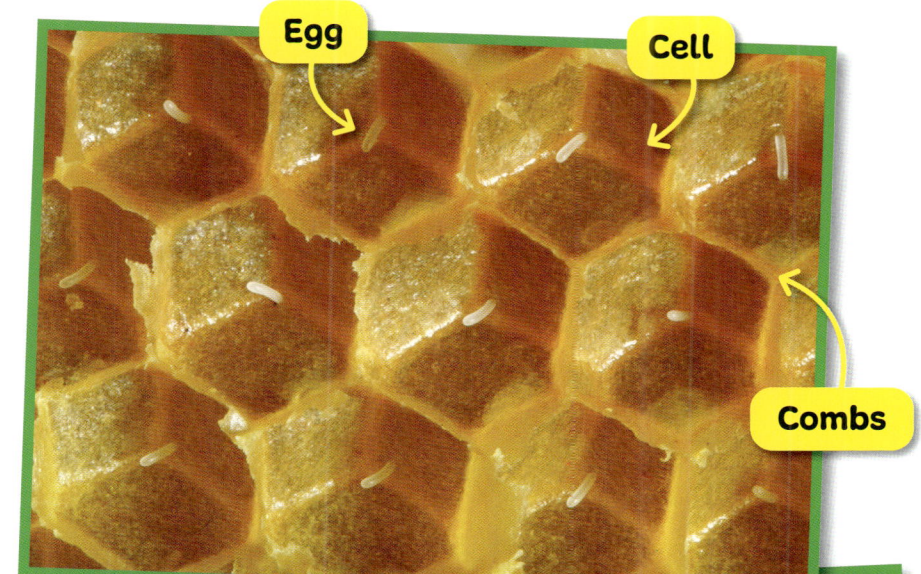

After three days, a baby bee called a larva hatches from each egg.

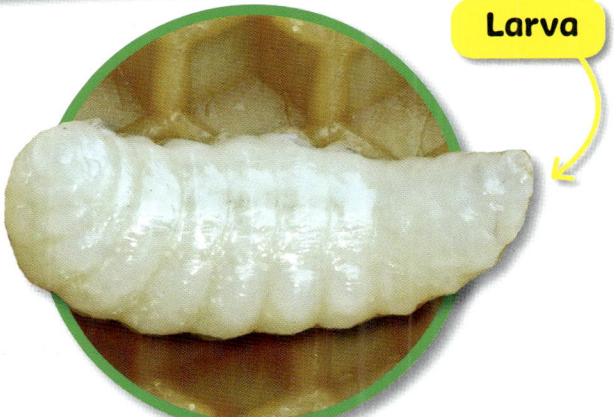

What Do Baby Honeybees Eat?

The worker bees visit flowers. They collect pollen and sweet, runny nectar.

Back at the hive, workers turn the pollen and nectar into bee bread.

The workers feed this special food to the larvae.

Worker honeybee

Pollen

A larva eats bee bread more than 1000 times a day!

Inside its cell, a larva becomes a **pupa**.

It takes about 10 days for the pupa to change into a honeybee.

Honeybee pupa

Worker

This new honeybee is climbing from its cell.

New honeybee

You will soon be visiting flowers, little honeybee!

How Many Eggs Does a Ladybird Lay?

Ladybirds are small flying beetles.

Ladybird

Like butterflies and bees, ladybirds are insects.

Female

Male

Male and female ladybirds meet up to mate.

After mating, a female ladybird lays her tiny yellow eggs under a leaf.

She can lay up to 300 eggs.

A female ladybird doesn't take care of her eggs.

Eggs

About 10 days later, a baby ladybird called a larva hatches from each egg.

Empty eggs

Larva

What happens next?

How Do Baby Ladybirds Become Red and Spotty?

A hungry ladybird larva starts feeding on tiny insects called aphids.

Aphid

Larva

After about a month, the larva's skin splits.

It has become a pupa. Now a lot more changes happen.

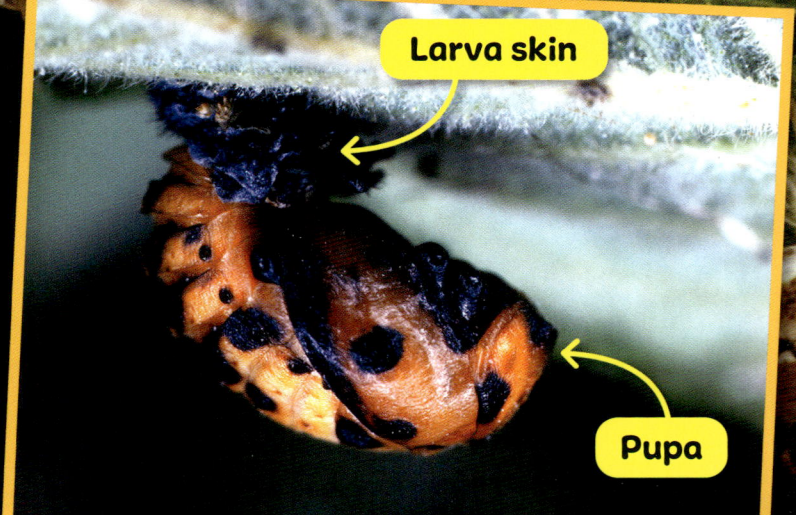

Larva skin

Pupa

About two weeks later, the pupa's skin, or case, splits open.

A yellow beetle climbs from the case.

Beetle

Pupa case

Now something amazing happens.

In just a few hours, the beetle turns red and grows black spots.

It's a new ladybird!

What Goes Around and Around?

It's a life cycle!
We can look at an animal's life cycle as a circle.

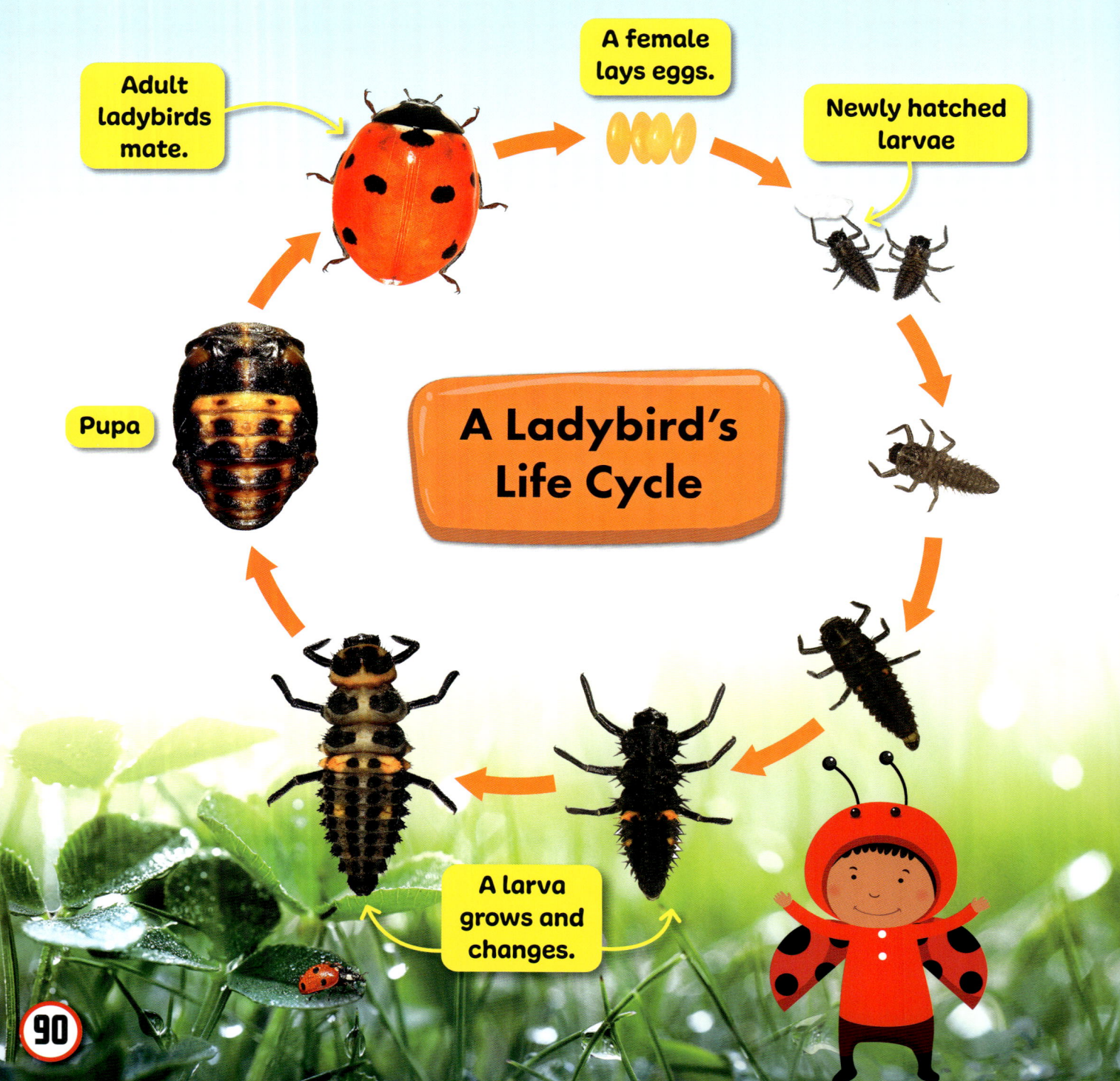

A Ladybird's Life Cycle

- A female lays eggs.
- Newly hatched larvae
- A larva grows and changes.
- Pupa
- Adult ladybirds mate.

Around, and Around, and Around!

Look at this duck's life cycle. We can see what's happening inside the egg!

A Duck's Life Cycle

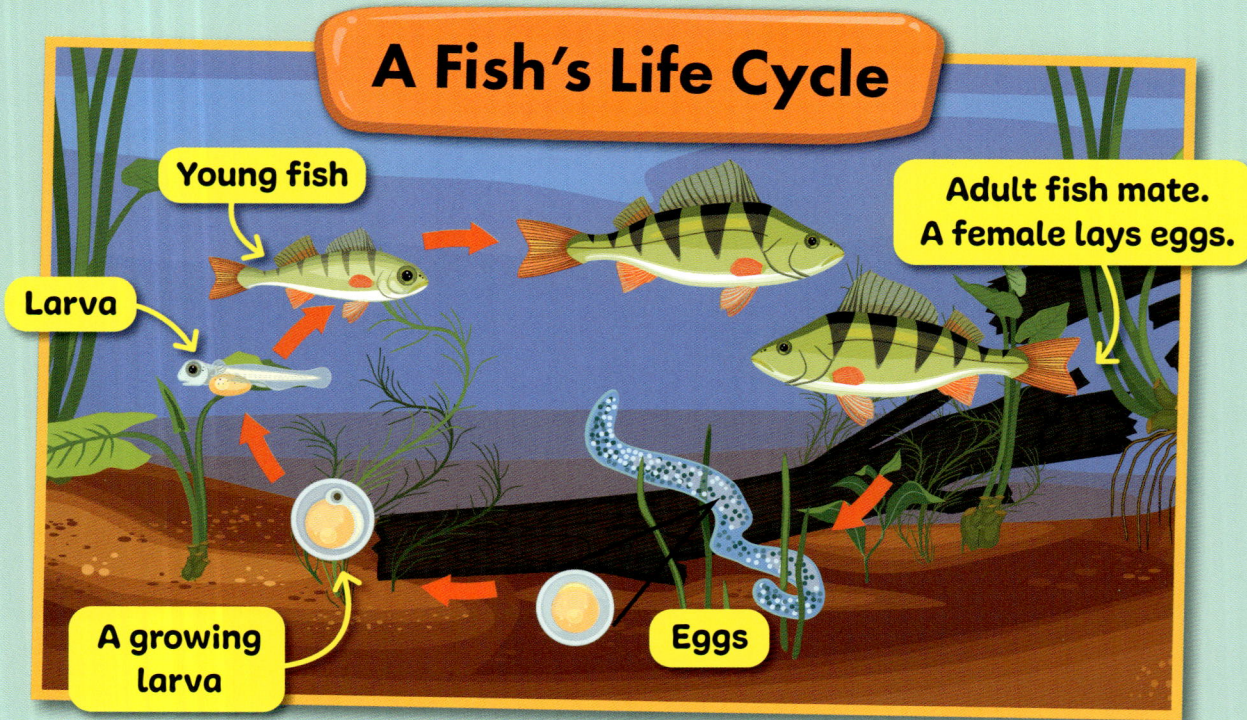

A Fish's Life Cycle

My Life Cycle Words

amphibians
Animals such as frogs and toads. Most young amphibians hatch from eggs, live in water and breathe with gills. Adults live on land and in water. Most adult amphibians breathe air.

birds
Animals with feathers, wings and a beak. All female birds lay eggs. Most birds can fly.

burrow
A hole or tunnel dug by an animal as its home.

den
A type of animal home. A den may often be hidden.

desert
A place where very little rain falls. Most deserts are hot and dry with sandy, rocky ground.

fish
Animals that live in water and have scaly skin, gills for breathing and fins. Most female fish lay eggs, but some give birth to live babies.

gills
Body parts that some animals, such as fish and frog tadpoles, use for breathing. Gills take in oxygen from water.

insects
Small animals with six legs and an outer shell called an exoskeleton. Most female insects lay eggs.

larvae
The young of some animals, such as insects and amphibians. Many larvae have long, fat bodies.

lungs
Body parts for breathing air.

mammals
Animals with hair or fur. Female mammals give birth to live babies and feed them milk.

mate
To come together to produce young.

predator
An animal that hunts other animals for food.

pupa
The stage in the life cycle of some insects when they change from larvae into adults.

reproduce
To make more of something. When animals reproduce, they make more of themselves!

reptiles
Animals such as snakes, lizards, crocodiles and alligators that have scaly skin. Most female reptiles lay eggs, but some give birth to live babies.

scaly
Having skin that's covered with small, tough parts called scales.

spiders
Small animals with eight legs and an outer shell called an exoskeleton. All baby spiders hatch from eggs.

Big Life Cycles Quiz

1: What is a life cycle?
  a) A type of exercise bike
  b) The stages in a life
  c) When an animal becomes an adult

2: What animals feed their babies milk?
  a) Reptiles
  b) Birds
  c) Mammals

3: Which baby animal is the biggest of all?
  a) An elephant calf
  b) A blue whale calf
  c) A giant tortoise baby

4: Which bird lays the biggest egg?
  a) An ostrich
  b) An emperor penguin
  c) A blackbird

5: How many eggs does a frog lay?
  a) 100,000 eggs
  b) 3 eggs
  c) 2000 eggs

6: What is a baby toad called?
  a) A chick
  b) A tadpole
  c) A hatchling

7: Where does a sea turtle lay her eggs?
  a) On a sandy beach
  b) In a tree hole
  c) In a puddle

8: How does a crocodile know her babies are hatching?
  a) They send her a text
  b) They call to her
  c) She hears the eggs cracking

9: Which animal mum might eat dad?
  a) A tiger mum
  b) A ladybird mum
  c) A spider mum

10: What is a pupa?
  a) A stage in an insect's life cycle
  b) A baby fish
  c) When a baby animal goes to the toilet

Answers:
1) b 2) c 3) b 4) a 5) c 6) b 7) a 8) c 9) b 10) a